Comets

Their appearance and significance

Georg Blattmann

Comets

Their appearance and significance

Floris Books

Translated by Donald Maclean
Edited by Christian Maclean

First published in German under the title
Das Rätsel der Kometen
by Verlag Urachhaus, Stuttgart, 1974.
First published in English in 1985 by Floris Books.

British Library Cataloguing in Publication Data

Blattman, Georg
Comets: their appearance and significance.
1. Comets
I. Title II. Das Rätsel der Kometen. *English*
523.6 QB721

ISBN 0-86315-016-0

Printed in Great Britain
at the University Press, Oxford

Contents

1. A comet's appearance and 'biography'

The most famous comet, Halley's, is making one of its periodic visits during 1985/86. The last time it was near the earth and could be seen with the naked eye was in 1910. Then its path was observed so closely and calculated so accurately, that on October 16, 1982 two astronomers at Palomar Observatory, California discovered the comet between the constellations of the Lesser Dog and the Twins, at about the same distance as Saturn is from the earth. Never before has a comet been spotted at this distance, and we can but admire the accuracy of the observations and calculations which have led to such a result.

However, while the path of a comet is quite predictable, no one can say whether Halley's Comet will reach such brilliance as to be visible during the day as was the case in May 1910. Certainly the fact that it can already be observed with powerful telescopes at the distance of Saturn appears to promise a brilliant object; but we have only to cast our minds back a few years to 1973 when Dr Kohoutek discovered a comet approaching at the distance of Jupiter. Great expectations were raised and newspaper articles and books appeared with such grand and, as it turned out, misleading titles as 'Kohoutek — Comet of the Century'. Dr Kohoutek had warned on discovery, 'No astronomer can already predict exactly what the maximum brilliance of this comet will be, or what size and kind of tail it will develop.' Comet Kohoutek did not live up to expectations: it never reached great brilliancy, and at its brightest could barely be seen with the naked eye. Today no one even remembers it.

We see that there is an unpredictable element in the appearance of comets. Patrick Moore in his book *Halley's Comet* calls them 'eccentric visitors from space', and we might add, that they are even wilful in their behaviour. They are different from other celestial bodies in that they show a certain freedom of movement. An astronomer at the beginning of the century once said that the main feature of the comets was that they did not obey the laws of the rest of the solar system. We can calculate accurately the positions of the moon, planets and the fixed stars, and foretell their movements in the years, decades and centuries to come, but we cannot do that with comets. The comets really have a certain element of freedom. Now journalists are inventive people and they have been searching for suitable words to describe such a remarkable entity, and so in one newspaper we could read about 'the vagabond of the sky'. That is not at all a bad expression, for it really does convey something of the roaming, unpredictable nature of the comet. But instead of calling them vagabonds we might perhaps be able to find a more friendly and accurate expression: we could do better to call them 'the freedom heroes of the universe'.

Here it will not be out of place to interpose the following thought: there could never be any freedom in the affairs and destinies of mankind, were not freedom already manifest somewhere in the cosmos; and that place of freedom is none other than the realm of the comets. Here we come upon a most important point: the connection of the comets to man.

1. A COMET'S APPEARANCE

Usually we think the comets so strange, so alien and so remote from our world that we are uncannily fascinated by them. It is so rare an occurrence when a comet crosses the sky that it causes a genuine sensation. Nevertheless we should recognize that in their intrinsic nature comets are closely connected to us human beings. If we observe them with sufficient attention and application we discover in them more and more features which are very much related to our own nature. We have already mentioned the matter of freedom. Freedom plays a significant part in our view of life and existence, even though in the main our destiny as human beings is preordained and we cannot escape from it, for we must all be born, pass through an ascendant phase of life, reach a certain culmination, followed by a decline terminating for us all in death, and in this respect our destiny is preordained and we cannot break this pattern. But within this far-stretched framework individual destiny is variable and can be shaped in the highest degree. That is where the freedom lies, and it is the same with comets. We can sketch the main features of a comet's biography. I deliberately use the word 'biography' although of course in astronomy it would hardly be accepted. Normally we do not speak of the biography of a celestial body. The sun has no biography, nor has the moon, Mercury or Mars, and as for the fixed stars, why, they have absolutely no biography at all, but the comet really has something like a life story. One day it is born, tiny and insignificant (later we shall examine this more closely and in detail); then it has an ascendant period until it reaches its greatest brilliance followed by a corresponding decline and 'death', or at least its complete disappearance from view. But within this 'far-stretched' frame the course of the comet's life is tremendously variable, differentiated, and in every phase unpredictable.

But this *visible* part of the comet's life is only a fraction of its total existence, for after the comet's 'death' it vanishes from our sight. After a periodic lapse of time the comet reappears, is 'reborn' as it were. Astronomers are quite accustomed to the idea of the reappearance of comets — indeed it was Halley himself who first propounded the idea — and are able to predict their return. Nevertheless when we apply this thought of death and rebirth to human existence we find that people do not always accept this thought quite so readily, and yet a true study of human biography should lead us quite naturally to accept it.

Now I propose to elaborate further this relationship of the comet with the human being. The ancients saw the comets connected with those souls that were leaving the earth. In a wonderful passage Ovid describes how the soul of Caesar after his murder ascends into the heavenly spheres, the goddess Venus bearing him upwards stage by stage into the heavenly world. There we see the soul of Caesar pictured as a comet 'drawing its flaming hair after it'. It is from this idea that 'comet' gets its name, being derived from the Greek αστὴρ κομήτης (*astèr kométés*), the long-haired star.

Not uncommon then is the idea that the comet is like a human soul making its way up into the realm of the stars. Another analogy with the comet is made when we speak of the 'meteoric' rise to fame of some brilliant young man. Here again we should note that just as there are innumerable less significant people coexistent with famous individuals so there are myriads of lesser comets besides the great and dazzling ones. Many laymen do not realize how many there are. Kepler has said that they are as numerous as the fishes in the sea, and modern astronomy confirms this, reckoning that in all there are a hundred thousand million, a quantity defying imagination.

Astronomers speak of 'a comet cloud', which lies round the outside of our planetary system and contains these millions of comets in a kind of dormant condition. It is a realm of unlimited cometary possibilities, from which a single or several comets are sent out from time to time to enter our planetary system in order to accomplish a particular task or mission. Thus a comet is born, making its appearance to us humans. It runs its course, carries out its mission and vanishes again into the comet cloud, there to wait patiently until it is allotted a new task.

We might compare this comet cloud and all that it suggests with the cloud formation of that famous work of art, Raphael's *Sistine Madonna*. There Mary stands with the Child on her arm in the middle of the picture with clouds in the background, and this background of clouds consists of many little faces of children and angels. That is 'the cloud of the unborn' around the figure of the Madonna. Out of this cloud of unborn souls awaiting incarnation Mary brings one special child into the world that he may carry out his earthly mission while the others remain in the background awaiting their call, each to his own hour of birth and earthly task.

Not without cause have painters represented as a comet the star which led the magi to the Adoration of the Child, even though today we are of the opinion that it was not a comet but a certain constellation which guided the Kings. This is not the place to go into this question further, for it is a whole subject in itself and has been treated in detail by Dr Walther Bühler.* Nevertheless the comet is the proper sign for indicating that a soul with a special mission is about to enter the earthly world, and so the inspired painters have set a comet above the birthplace of Jesus to indicate the Incarnation.

The idea that a comet also has a mission is even to be found among journalists, and one wrote, 'Comets, messengers from space'. Although that was a stroke of genius I do not think that the journalist really knew the truth of what he wrote. But I do believe that the comets really are messengers, special messengers sent upon their special missions.

We have now seen that whereas in earlier times man's gaze was directed more to the end of life when the departing soul was seen in its connection and likeness with a comet disappearing, we have taken the aspect of the comet coming from the realm before birth.

* 'Die grosse Konjunktion', *Sternkalender 1960/61*, Philosophisch-Anthroposophischer Verlag, Dornach, Switzerland.

2. The task of the comet

Let us now look more closely at the kind of task which a comet may have in our world. People in the Middle Ages watched the appearance of comets with great fear and apprehension, and even now who would proceed unmoved and indifferent, if on his walk he should chance to look up and suddenly see a large comet? In our lifetime there have not been many occasions when we could see a large comet. The oldest may remember the Great Daylight Comet of January and February 1910, and days later Halley's Comet which by May of that year could be clearly seen in the eastern twilight before dawn. My best observation of a comet was in the spring of 1960 when Bennett's Comet was visible in the eastern sky in the early morning. In the fifties and sixties Comet Arend-Roland and Comet Mrkós were also to be seen, but although none of these was particularly exciting we were glad enough to have seen a star with a tail. In the accounts of former times however we can read the most astonishing things. A German astronomer in the observatory in Athens* described how he went out one evening just as darkness was falling and suddenly saw a comet with its head on the horizon and its tail reaching to the zenith. Even the astronomer got a fright!

But let us return to the people of the Middle Ages

* Quoted by K. Wurm in *Die Kometen*, Springer, Berlin 1954.

who watched the phenomena of the skies with much more childlike, naïve and natural feelings than we over-educated people of the twentieth century do. Comets gave them a most uncomfortable feeling. They would say that the comets were the scourge of God, were the sign of divine displeasure calling upon men to change their ways. Comets were portents of the worst disasters: wars, pestilence and plagues, foul weather and so on. It was like at school when a teacher lets loose a storm of anger on his class. That is most unpleasant for his pupils at the time — and so it should be, otherwise it would not have the desired effect. His pupils are filled with fear and apprehension, but once the storm is over and everyone has had a bit of a shock, and order has been restored, then the good and salutary effect of such an outburst can be felt. The same can be said about a thunderstorm. There is hardly anyone who looks forward to a thunderstorm with joy or satisfaction. When the horizon grows black, when lightning begins to flash and thunder rolls and echoes round the mountains, who does not feel at least some qualms in his inmost being? But once the thunderstorm is over the benefit is felt. The air and all around is clean and fresh not only from the rain but as a result of the tremendous discharge and convulsion of the elements. Afterwards the world seems newly-created and we feel the countryside can begin to live afresh and with it all creatures. So we see the thunderstorm accompanied by an anxious moment before and while it is upon us and by the subsequent beneficent effects that bring the feeling of having been born anew.

In much the same way people felt the appearance of a comet. After it had appeared they felt the world cleansed and purified, as if the comet had been a celestial broom sweeping away the mess that had been accumulating in the universe. All life is renewed and can make a new beginning.

Let us take another example, that of a fever going through the body. It goes as a wave of heat, linked perhaps with pain. There is a condition known as 'hot flushes', passing from one part of the body to another. No one who has suffered this wave of heat and pain likes it, indeed we tend to send for the doctor and try to get rid of it as soon as possible. But once the fever has left the body we often feel quite refreshed. People say: 'after the fever I feel reborn, still rather weak, but relieved and lightened'. The fever by raising the temperature has loosened and swept away all kinds of deposits and residues, and so life can make a fresh start. In much the same way a comet passes through the organism of our planetary system as a kind of transient heat wave, a roaming inflammation, a cosmic thunderstorm, leaving behind a cleansed, purified living organism.

Now let us look at those comets which reappear, or which are at least assumed to reappear. In the case of a number of comets it has been possible to calculate exactly when they are due to reappear, although there are problems here because, as I have already said, every comet disappears completely from sight, and we cannot stick a tag on it beforehand so that it can be identified when it comes back. No, it vanishes completely and reconstructs itself. The reconstructed comet now may move on a path so similar to the original comet's, and can appear so similar to it in every way, that it is quite natural to assume that the one which vanished is the same as the one that has reappeared. But let us leave aside this thorny question of whether it is indeed the same comet or in how far it is the same. Let us simply take what the astronomers say: there are comets which return.

Edmund Halley (1656–1742) who first postulated that comets return was a contemporary of Isaac Newton, Robert Hooke and Christopher Wren. He succeeded John Flamsteed as Astronomer Royal in 1719. In 1682 a bright comet was discovered and astronomers observed and measured its course. Halley made calculations of the orbit and was struck by the remarkable similarity to comets previously observed in 1607 and 1531. Checking further back in time Halley came across comets seen in 1456, 1378 and 1301 which showed similarities. He felt confident enough to publish his results forecasting a return of the comet around 1758. Though Halley did not live to see it, the comet was duly spotted at Christmas 1758.

Following back this discovery of a 76-year recurrence in history it can be traced by regular reports through 1066 (it is depicted on the Bayeux Tapestry of the Battle of Hastings), to 12 BC. These reports are particularly from China, but also the Jewish historian Flavius Josephus mentions a comet 'that stayed a whole year' at the siege of Jerusalem in AD 66. There appears no record of its visit in 164 BC, but Chinese records describe a bright comet in 240 BC, that is twenty-nine 76-year periods before its current appearance.

While Halley's Comet is undoubtedly the most famous, there are a number of comets whose periods range from as short as 3.3 years to about 200 years. Theoretically the periods can be much longer, but the probable error in measurements prevents accurate periods being calculated from these comets, and also they have little practical value.

2. THE TASK OF THE COMET

What is the significance of their continuous returning? I should like to illustrate that by an analogy. Supposing someone cleans a house. That is always very commendable if the house is very dirty and needs it. But cleaning can go too far, and if a housewife always has a brush in her hand and is always sweeping when there is not any dirt about, then her cleaning becomes a pain in the neck for everyone. Then we say that that person is possessed by a demon of cleanliness or that she is one herself. So we see that an intrinsically positive activity can become negative if it is overdone. Now it would seem that this is what happens with comets, so that a comet which has been sent out with a particular task, after completing it, cannot leave off. It has to go on repeating the thing even though the necessity for it has long been absolved. So what was originally a positive task now just becomes a stereotyped habit, and thus pathological. Accordingly we have the distinct impression that some of the comets are suffering from a kind of demonic possession which drives them to repeat the same thing over and over again, and what was originally intended as a benefit to the whole solar system now becomes a plague.

There is clearly a high authority, a cosmic authority which sends out comets to carry out certain missions; so too there is an authority which can put a stop to the comets which have run amok. That is not just a simple matter of putting up a stop sign, for when such a possessed devil is careering about it is not enough simply to shout 'halt'. It is like the magician's apprentice: once the broom has started sweeping it is not so easy to make it stop. But there is a watchman in the universe who will have to wield the truncheon and smash the wild fellow to bits. In his wonderful poem

'Die Schlacht am Himmel' (The Battle in the Sky)* Albert Steffen describes how Biela's Comet, which was one of those in the last century which could not stop, was broken up in a mighty cosmic battle. First the comet was divided into two and later it was split up into innumerable little pieces which appeared as a tremendous shower of shooting stars. This meteor shower appeared first in 1872 and Rudolf Steiner saw it as an eleven-year-old boy. The stars rained down from the sky in such numbers that people were alarmed and thought that the sky would lose all its stars. This was repeated a few times but not with such intensity. Only once did it rain again so fiercely and some people may still remember it — that was on October 10, 1933. Since then this phenomenon has been completely transformed and taken up into quite a different rhythm. It can be seen to some extent every year but is by no means so impressive and spectacular.

The breaking up of Biela's Comet marked a significant turning point in cometary history. During the nineteenth century there was quite a witches' sabbath of comets with quite incredible phenomena, and going back in time to the Middle Ages we have many impressive accounts of such phenomena. There were comets with tails reaching over half or even all the sky; there were comets with as many as ten tails and in the most amazing forms. All this activity and spookiness suddenly ceased at the turn of the century, soon after the breaking up of Biela's Comet, and one is aware that a new period of history — cometary and human — has begun. In our century the sky has been relatively empty of comets, and so the return of Halley's Comet raises

* In *Wegzehrung*, Verlag für Schöne Wissenschaften, Dornach 1935.

great expectations. Hardly any of us has ever seen a really big and splendid comet and so everyone is looking forward to seeing one. Our previous expectations of sighted comets were not satisfied, for living up to their erratic, eccentric behaviour, they did not reach the brilliancy forecasted by sensation-seekers.

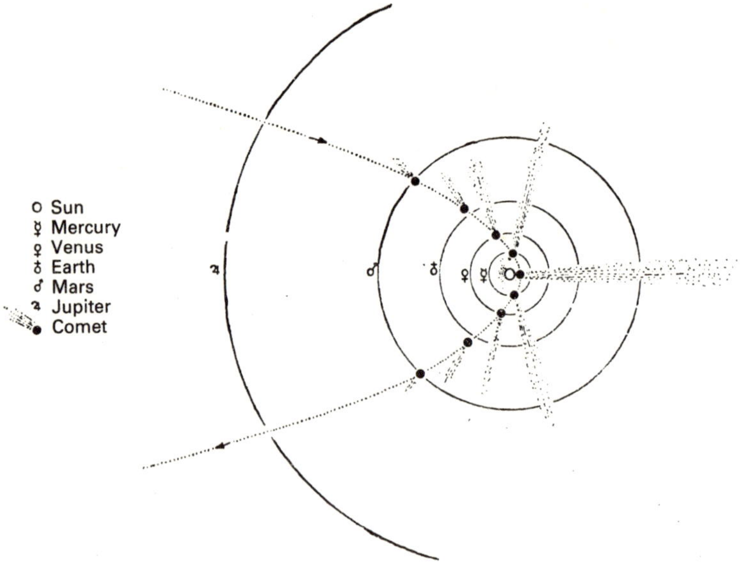

3. The tail of a comet

Now let us take the forgotten Kohoutek as an example in order to describe the life story of a comet in greater detail. It was Dr Kohoutek's job to watch the skies every night and to photograph them — nowadays a camera is used and so you can lay the photographs conveniently on the table and compare them at your ease to see if there have been any changes anywhere or if anything unexpected has turned up. One day in March 1973 Dr Kohoutek discovered a little bright spot, a small misty insignificant spot, where nothing had been before. He spotted it right away and then he examined each new day's photographs very carefully and saw how this spot moved daily and grew in size and brightness. Gradually it began to grow a tail just like a tadpole, and the further it went the bigger and brighter its tail became. Dr Kohoutek of course knew by now that it was a comet.

If we imagine the sun as being in the centre of our solar system and the planets moving around it in various elliptical paths (see figure), the comet enters on an elongated elliptical path while the planets' paths, though also elliptical, are almost circular. The comet crosses the planets' orbits, comes quite near to the sun while its tail becomes bigger and bigger. The tail reaches its greatest size when the comet is nearest the sun. The comet now makes its exit on a path symmetrical with that of its entry, and the tail becomes smaller until it is only a patch of light and finally disappears altogether. Such is roughly the life story of a comet.

Generally comets become visible as they cross the orbit of Mars and become invisible when they cross it again. That is the rough rule. A few comets, however, become visible already as they cross the orbit of Jupiter much further away, and Halley's Comet, as was mentioned earlier was discovered at the distance of Saturn.

The figure shows a remarkable fact which the reader may have noticed, namely the direction of the tail. The comet does not draw its tail after it as one might be inclined to think, or as one might infer from Ovid when he speaks of the comet 'drawing the hair of his head after him,' but the tail always points away from the sun. As the comet approaches the sun the tail follows, then the comet sweeps its tail round the sun and it follows it on its outward path. In this way the tail sweeps round in a much wider path than the head. And if we take into consideration Kepler's Law that the comet moves at its fastest when it is near the sun, and that the tail is at its longest just at this sharp bend, we can see at what a tremendous speed the end of the tail must be travelling. Depending on its distance from the sun the head itself can attain the incredible speed of up to 400 miles per second (650 km/s). The tail is several millions of miles long, so what sort of speed does it reach? That is a question that has troubled astronomers for long enough, and still they have not come up with a satisfactory answer. Moreover they have found out that the tail is extremely thin and fine, not like a stick which you can take in your hand and swish through the air. Even a stick or metal rod would bend if it were to be whipped through the air quickly. But the comet's tail is thinner than a film of mist, for when it passes in front of a star, however faint the star is, the star will shine through the tail. Even when the tail passes in front of the sun it in no way darkens it as you might expect. Yet in spite of its tenuousness it is whisked about out there in space.

The comet's tail is something like a searchlight flashing its beam into the sky and moving its finger across the sky in search of its target. The beam moves according to the direction in which the lamp is pointed and does not bend however swift the movement of the lamp. The reason why the beam does not bend is because it does not consist of matter. If the weather is misty the finger of light can be seen all the better. It is in fact not the light which one sees but the lit up vapour. When the beam moves it is not matter which moves but the light situation, something which is non-material, and because it is not matter which moves, there is no difficulty about the speed of the movement.

A French astronomer invented a wonderful figure of

speech to describe the comets. He called them *des riens visibles* (visible nothings). Like the beam of the searchlight the comet can move at any speed through space without opposition.

Now let us apply this knowledge within a greater framework. Let us apply it to what I have already demonstrated concerning the structure of our solar universe in my book on the sun.* In that book I mentioned the sun's corona which can be observed during an eclipse. This corona reaches out in all directions, and this mantle of rays is so vast — and this has been discovered by the latest research (especially from the results of satellites) — that our whole planetary system is immersed in it; and the planets so to speak swim along their orbits in a sea of light whose 'waters' (a very fine medium) are not still but are agitated in strongly pulsing storm waves.

Now let us follow the path of the comet once more. The comet moves in the ocean of the corona. From the periphery it swims to the middle. As a kind of centre of disturbance the head passes through the outstreaming of the corona. As it moves forward, cutting the lines of the streams, something that formerly was invisible now becomes visible. What is that something? The comet's tail becomes more and more clearly defined the nearer the head gets to its perihelion (the point at which it is nearest the sun). The tail shows a fibrous raylike structure very similar to that of the streaming field of the corona.† The fanlike bundle of rays that attach them-

* *Die Sonne — Gestirn und Gottheit*, Urachhaus, Stuttgart 1972. English translation to appear 1985/86 Floris Books.

† Not only can the structure be compared but also the speeds registered in the comet's tail and in the flowing sun plasma of interplanetary space show correspondences.

selves as a tail to the comet's head are nothing less than parts of the great garland of coronal rays that become successively visible as the head moves along. What was invisible now becomes visible in the same way as the searchlight lights up what formerly was invisible in the air or mist. The area made visible is what we see as the comet's tail. It moves along always pointing away from the sun. Going back to our physiological example we might say it is a kind of feverish inflammation that is running about in the sun's corona and makes different parts in succession perceptible, just as shooting pains or heat spasms make us feel different parts of our body, but if we are healthy then we do not feel our bodies at all.

What is the relevance of these 'visible nothings' and this empty condition which moves around, while we can read in the papers and in books on astronomy that the comets really do consist of something? We know exactly what it is, for the astronomers can tell us the different chemical elements of which the comets consist. That however does not really contradict what I have been saying. If the pain was first in my hand, then in my elbow and later in my shoulder-joint, then something was there, and a doctor can make tests and analyse exactly the chemical composition of the cells there. These have probably undergone change through the inflammation passing through them, and that also can be ascertained exactly; nevertheless it would be utter nonsense to say that the pain consisted of these chemical elements. It is the inflamed condition which makes perceptible what is there already, and it is the same with the comet. The comet is simply a condition which makes visible what was there already.

4. The polarity of moon and comet

It is amazing how poets may sometimes reach out and grasp something which science laboriously works out bit by bit decades later. The poet Börries von Münchhausen calls the comet 'born of the sun' and 'emerged from the sun'. The truth of these expressions is shown by the latest results of cometary astronomy, while the poet knew it long before by a stroke of genius.

The comet is really 'a bundle of sun-rays', a part of the sun. That brings me to the last part of what I have to say about the being or essence of the comet, namely to show this being in another light; or I might say, the comet itself can show us our whole universe in another light. When we contemplate the universe deeply enough we discover a fundamental polarity underlying our planetary system: the polarity of sun and earth. I have tried to illustrate this polarity from different angles in my book on the sun. The polarity shows the earth as a compact body of matter upon which we stand, and which contains within itself everything belonging to it. The sun on the other hand is of quite the opposite nature: its whole content is in its surroundings, in this mighty body of rays which we have just mentioned; and in its middle it is a kind of void, at whose edge this intense shining appears. To sum up, we have the earth as a compact mass of matter and the sun as a radiant being sending its rays to the widths of space with a shining pearl in the centre: that is the great fundamental polarity.

From this polarity we may observe extensions in both directions. When we intensify the quality of the earth as compacted matter in a dead, solid, impenetrable form we get an even deader, colder, even more solid mass and that is the moon. The moon is a piece of the earth ejected from the living earth body because it was too dead, too heavy to be carried along, too indigestible. Just as a stone, a kidney-stone, has to be ejected from the living organism, so the moon has been rejected from the living earth organism. It is really an excrement of the earth.

There is a corresponding intensification of the quality of the sun. This is produced by potentizing the sun's quality of outpouring, outgiving, outstreaming to infinity, and we get a bundle of sun-rays disseminated, scattered and lost, and that is the comet.

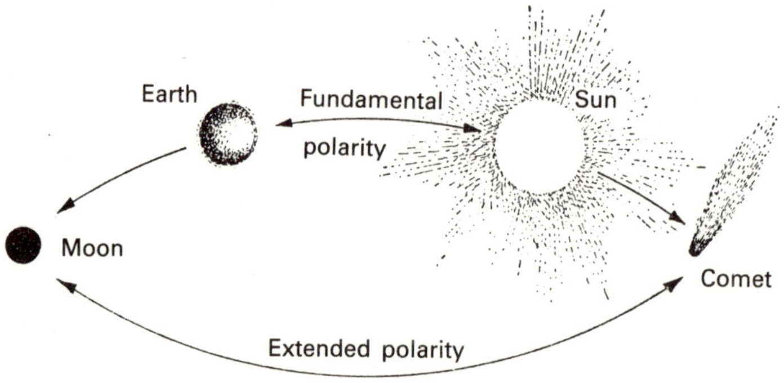

In the middle we have the fundamental polarity: sun and earth, and on each side the extended polarity: comet and moon; or better comet-being and moon-being, for in our planetary system there are several moons and as I have already said, a vast number of comets. We approach an understanding of the true nature of our planetary universe and its life when we contemplate the workings of this double polarity. With this key we may proceed further in unlocking much of what goes on behind the veils of the world, but I can here give only a few indications of what may be explored and studied.

Let us consider the plants for example. On one hand the life of the plant is determined by the forces of growth. These forces of growth are subject to the influence of the moon. On the other hand the future of the plant, and the food which the plant gives to man, are dependent upon another process which supersedes that of growth, namely the process of fruition. Fruition or ripening is the opposite process to growth. During ripening the growth gradually dies. Ripening is a gradual death. (And in human life, I might add, dying should be a ripening.) Now just as the moon is connected with the growth process, so the comet is connected with ripening. This was known by the great philosopher Hegel for he indicated that a specially good vintage followed significant appearances of comets. The quality of good wine is specially dependent upon the ripening process, because what is the use of large grapes if they are not sweet? Sweetness comes from the process opposite to growth, it comes not from the watery element but from the element of warmth and sunlight. Thus growth and ripening are two opposites that rule the world of life and are linked to the effects of moon and comet.

In the human realm we also have such a polarity: the male and the female. It makes all the difference whether one is a woman or a man, and this difference affects both body and soul. Our behaviour and our destiny also depend on whether we react as female or male. This differentiation is a reflection of the polarity of moon and comet. Once we have this key in our hand we can understand much in the destinies, encounters and liaisons of men and women, which otherwise would remain a mystery or at least unclear.

One other thing I should like to mention: the polarity of bread and wine is connected with the polarity of moon and comet. In the history of Christianity it has often been asked why Communion should be taken in the twofold form of bread and wine. Christ himself gave to his disciples this ceremony so that they might always be able to enter into a new communion with him: 'As often as ye do this, I shall be in your midst and work with you.' Would not one substance in which the Christ-being were present suffice? No, it would not suffice. To think that would be falling short in thought, for the universe is of twofold construction. Thus it cannot be otherwise than that the great mighty sun creative being, who in the Christ has joined himself with the earth, should impart himself in a twofold form. He is the Sun-God become an earthly man, and to those who follow him and seek his presence he comes in a twofold way, in bread and wine, in the one substance that is formed, baked into a loaf and cut out into the wonderful circular form of the host, and in the other that is a fluid without form, that moulds itself to the shape of whatever vessel contains it, but which has the property of being able in its liquidity and mobility to penetrate and enliven all things. There again we have the fundamental polarity.

Wherever we go in the world we meet these fundamental polarities. With this key in our hands the whole universe becomes great and wonderful for us, a landscape in which we can live as human beings, for in ourselves we find this polarity in what we think in our heads and what we do in our limbs. At first there may not be harmony between the opposites, but in the course of our lives we should learn to bring them under one rule; and because we as human beings are constructed as polarities we can feel at home in the universe. That is not however a matter of course, because we cannot really live in the universe portrayed by modern astronomy, a universe empty, desolate, cold, vast and uninviting.

That is why it is all the more important to attain a vision of the world which is not built upon this warped explanation but which is founded on real insights brought to us by the spearhead of research. We may then be grateful to Halley's Comet for giving us the opportunity and impulse to study astronomy more than ever before, and if the comet is not going to be as bright and magnificent as formerly at least he will show us how worthwhile it is to occupy ourselves with the secrets of the universe.

The Seven Planets

Adam Bittleston

Today's knowledge of the planets rests on scientific observation. Adam Bittleston takes the astronomical phenomena as his starting point, going on to embrace the intuitions of astronomers, poets and artists from varied cultures.

Bittleston brings diverse witnesses: the echoes of the Upanishads and the *Bhagavadgita*, the words of Buddha and Aristotle, the biblical sense for meaning in number, and insights of such minds as those of Kepler, Shelley and Goethe. He weaves these pieces together to give a picture of the planets showing something of their 'soul' and 'spirit'.

The various chapters gradually develop descriptions of the planets. The Moon, easiest to observe, progresses regularly around the zodiac in a miniature correspondence to the Sun. Describing lunar influences in the biosphere, Bittleston then approaches Rudolf Steiner's spiritual investigation, Anthroposophy. This allows consideration of a human existence before birth and after death, linked with the spheres of the planets.

Floris Books